The
Adventures of
Strawberryhead
& Gingerbread™
Fun Math!
For Kids ages 4-7

The Adventures of Strawberryhead & Gingerbread™ Fun Math!
For Kids ages 4-7

KF Wheatie & KM Wheatie

Strawberryhead & Gingerbread Press

www.strawberryheadandgingerbread.com

The Adventures of Strawberryhead & Gingerbread™,
Fun Math! For Kids ages 4-7

Published by Strawberryhead and Gingerbread Press
https://www.strawberryheadandgingerbread.com

ISBN: 979-8-9906129-7-6 (paperback)

Write the Numbers

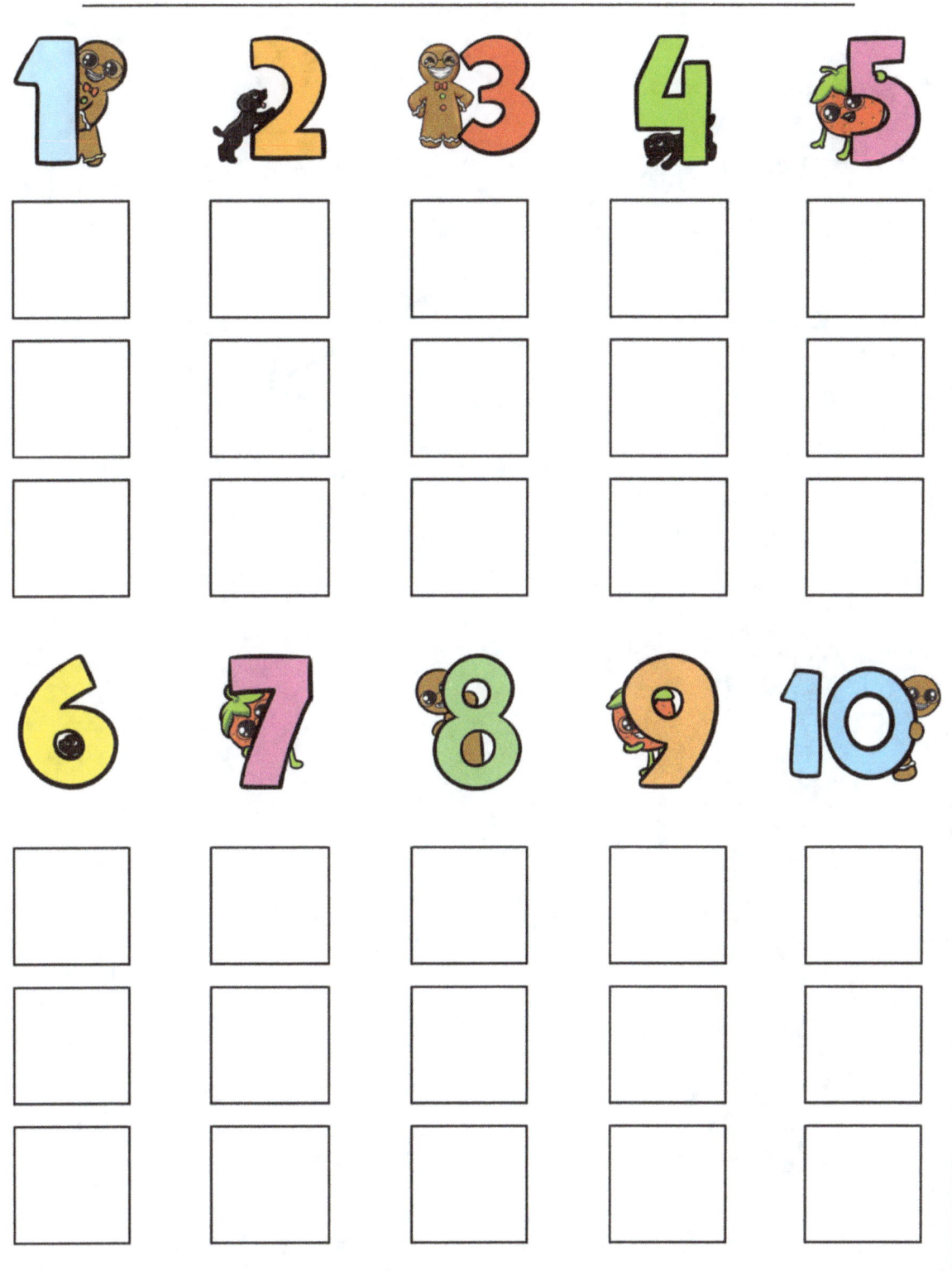

Match the Shapes

Count the Petals

_____

Circle the Big One

Write and Count

1	1	.	
2	2	.	
3	3	.	
4	4	.	
5	5	.	
6	6	.	
7	7	.	
8	8	.	
9	9	.	
10	10	.	

11	11	.	
12	12	.	
13	13	.	
14	14	.	
15	15	.	
16	16	.	
17	17	.	
18	18	.	
19	19	.	
20	20	.	

Count and Write

_____

   +   = ☐

 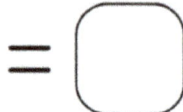 = ☐

  = ☐

 = ☐

  = ☐

 # Complete the Boxes

1		11	16
	7		
3	8		18
		14	
	10		20

Count and Compare

	9 11 8
	3 6 4
	7 5 6
	3 1 2
	9 7 5

Draw Shapes with Dots

Circle the Largest Number

(10)   4   6   3   1   10
1   10   8   2   9
1   2   9
    2   1   9   1   3   6
6   10   2   3   10   3
        1

11   12   16   11   10   20   11
            20   11   12   9   11
(20)   16
    2   1   6   20
        12   3
16   20   16   20   17

4   8   9   2   8   7   4   9
                7   3   7
7   2   6   2   9   2   7   2
3   (9)   7   4   7   6   9   4

Circle the Smallest Number

10   9   6   10   9   6   11

(6)   7   8   11   8

6   10   7   6

    7   6   9   9

11

---

16   12   16   12   20   12

    20   14   17

(12)   17   20   12   16

20   12   20   14   16

    17

---

    4   1   1   4

(1)   2   3   7   3

4   6   6   2

  7   3   4   1

1   1   3   6

    4   7   6   1

Count and Write

Subtract and Solve

8 9	4 8	4 7
- 7 5	- 1 6	- 3 5
1 4		

7 9	4 7	3 9
- 6 7	- 1 7	- 2 6

9 7	4 8	6 5
- 6 5	- 3 1	- 5 4

Multiply and Color

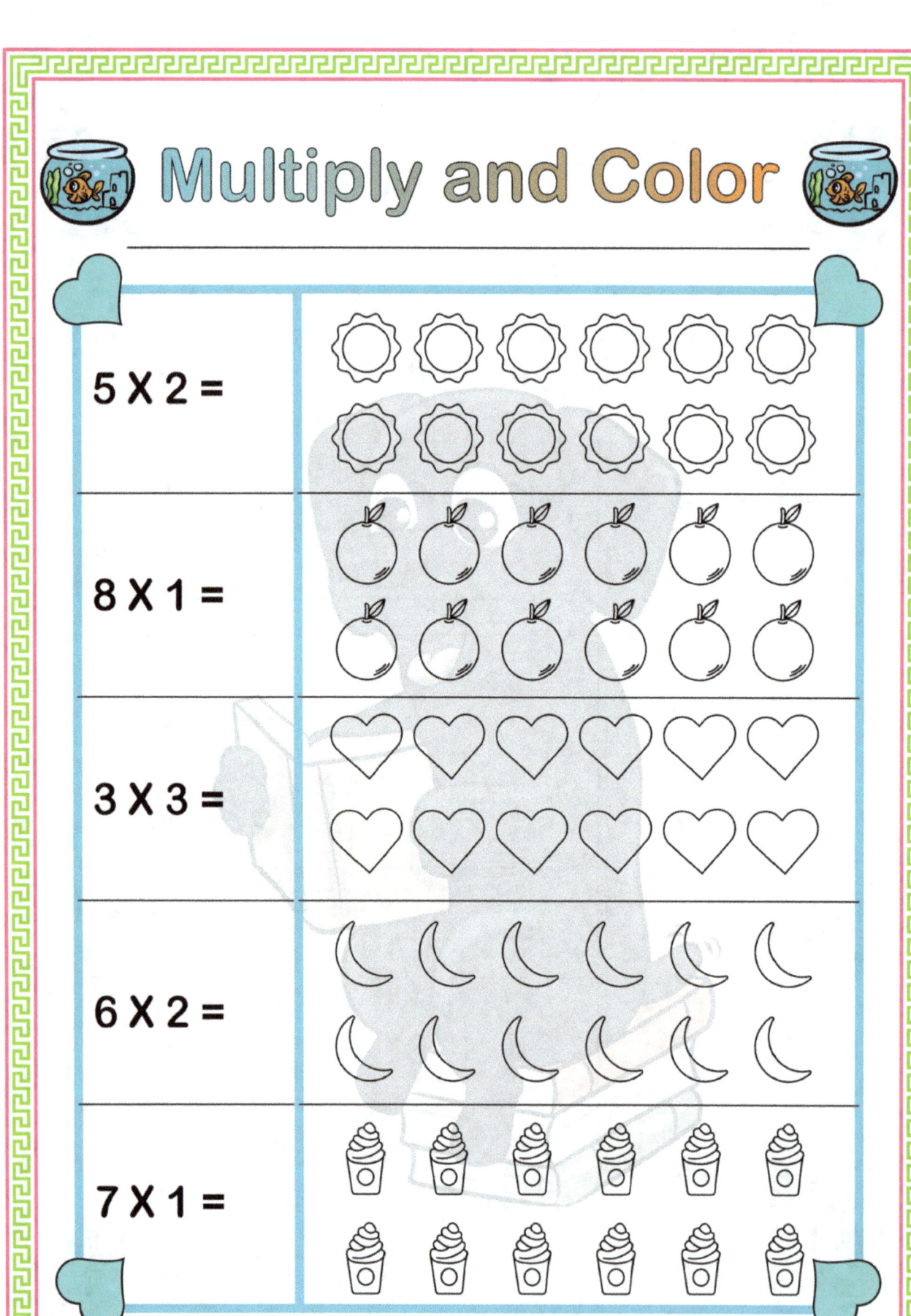

5 X 2 =

8 X 1 =

3 X 3 =

6 X 2 =

7 X 1 =

Addition Fun

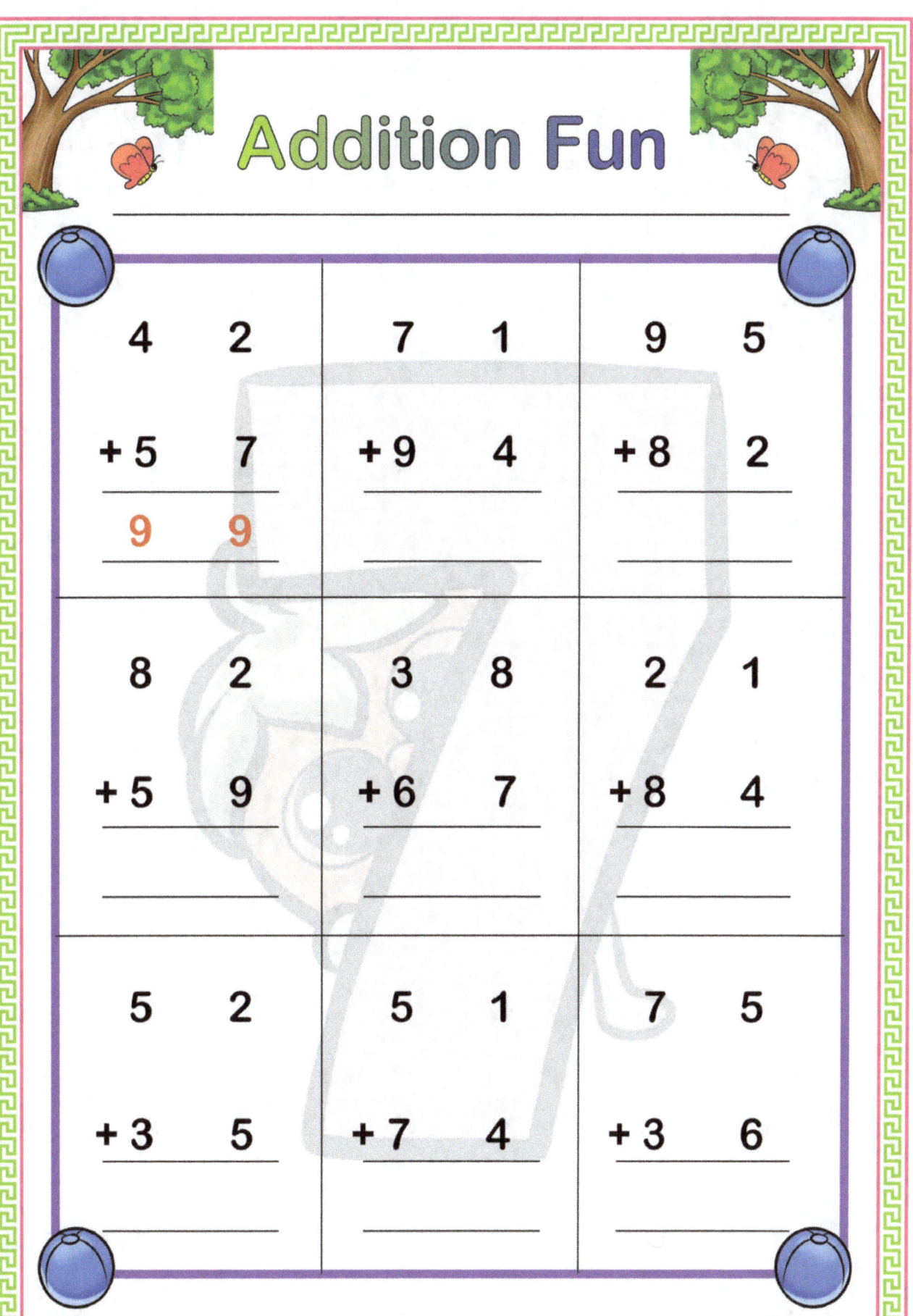

4	2	7	1	9	5
+ 5	7	+ 9	4	+ 8	2
9	9				

| 8 | 2 | 3 | 8 | 2 | 1 |
| + 5 | 9 | + 6 | 7 | + 8 | 4 |

| 5 | 2 | 5 | 1 | 7 | 5 |
| + 3 | 5 | + 7 | 4 | + 3 | 6 |

Multiplication Fun

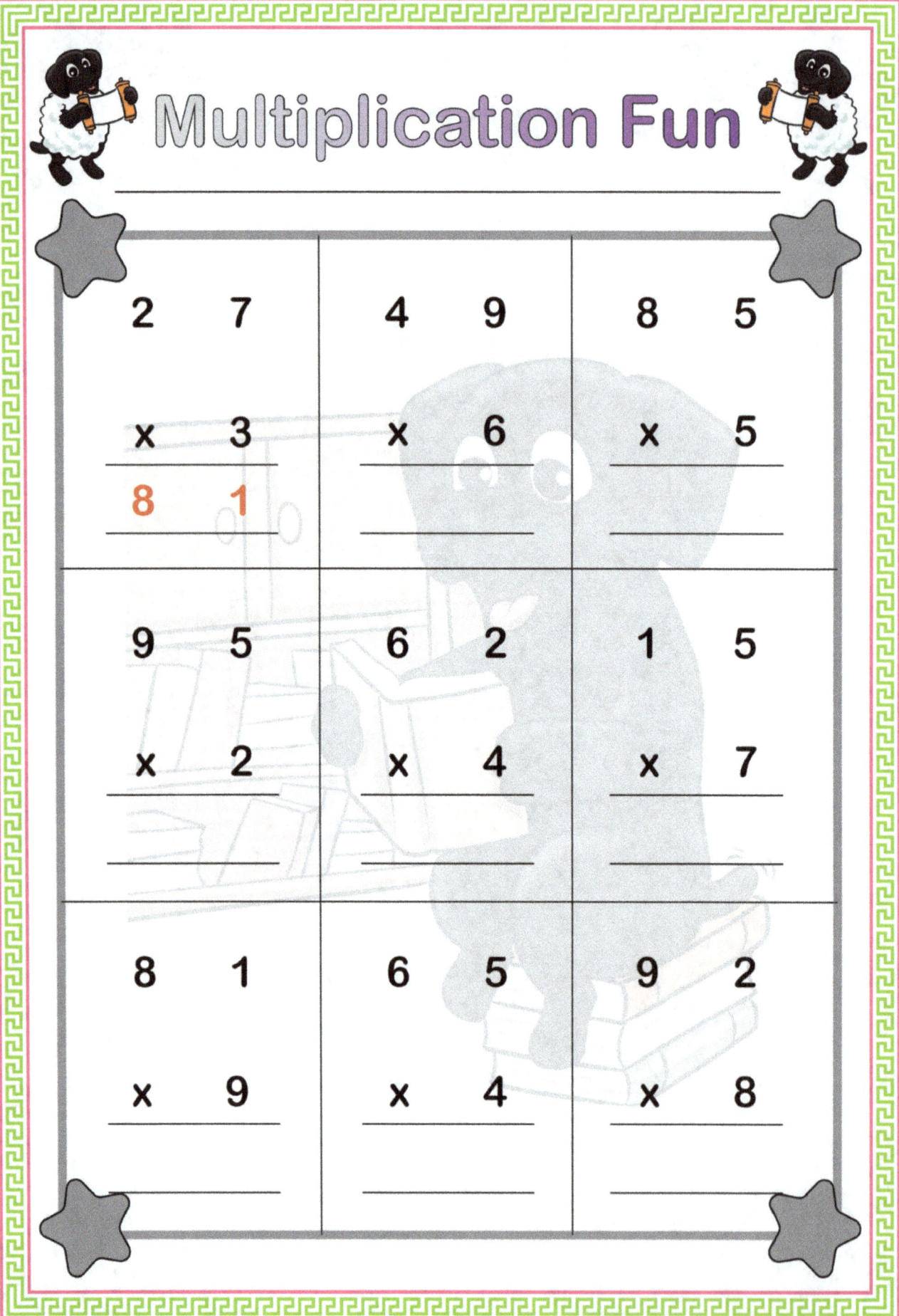

2	7	4	9	8	5
×	3	×	6	×	5
8	1				

9	5	6	2	1	5
×	2	×	4	×	7

8	1	6	5	9	2
×	9	×	4	×	8

Multiplication Fun

_____

2 x 9 =

2 x 2 =

2 x 1 =

2 x 4 =

2 x 3 =

2 x 6 =

2 x 5 =

2 x 8 =

2 x 7 =

2 x 10 =

 # Trace and write

1	One	One	1
2	Two	Two 2	
3	3 Three	Three	
4	Four	Four	4
5	Five	5 Five	
6	6 Six	Six	
7	Seven	Seven	7
8	Eight	Eight 8	
9	Nine 9	Nine	
10	Ten	Ten	10

 # Trace and write

11	Eleven	Eleven	
12	Twelve	Twelve	
13	Thirteen	Thirteen	
14	Fourteen	Fourteen	
15	Fifteen	Fifteen	
16	Sixteen	Sixteen	
17	Seventeen	Seventeen	
18	Eighteen	Eighteen	
19	Ninteen	Ninteen	
20	Twenty	Twenty	

Count the Balloons

 # Sort the Numbers

_____

Even Numbers	Odd Numbers

Fill in the Blanks

15	**7**	**11**	**20**
7 + 8	2 + ☐	6 + ☐	9 + ☐
3 + ☐	3 + ☐	9 + ☐	6 + ☐
9 + ☐	1 + ☐	7 + ☐	3 + ☐
10 + ☐	4 + ☐	2 + ☐	5 + ☐
14 + ☐	6 + ☐	4 + ☐	2 + ☐

Subtract and Solve

✋(4)	−	✋(4)	=	**1**
✋(5)	−	✋(3)	=	☐
✋(5)	−	✋(1)	=	☐
✋(3)	−	✋(3)	=	☐
✋(4)	−	✋(3)	=	☐
✋(3)	−	✋(1)	=	☐

Count and Write

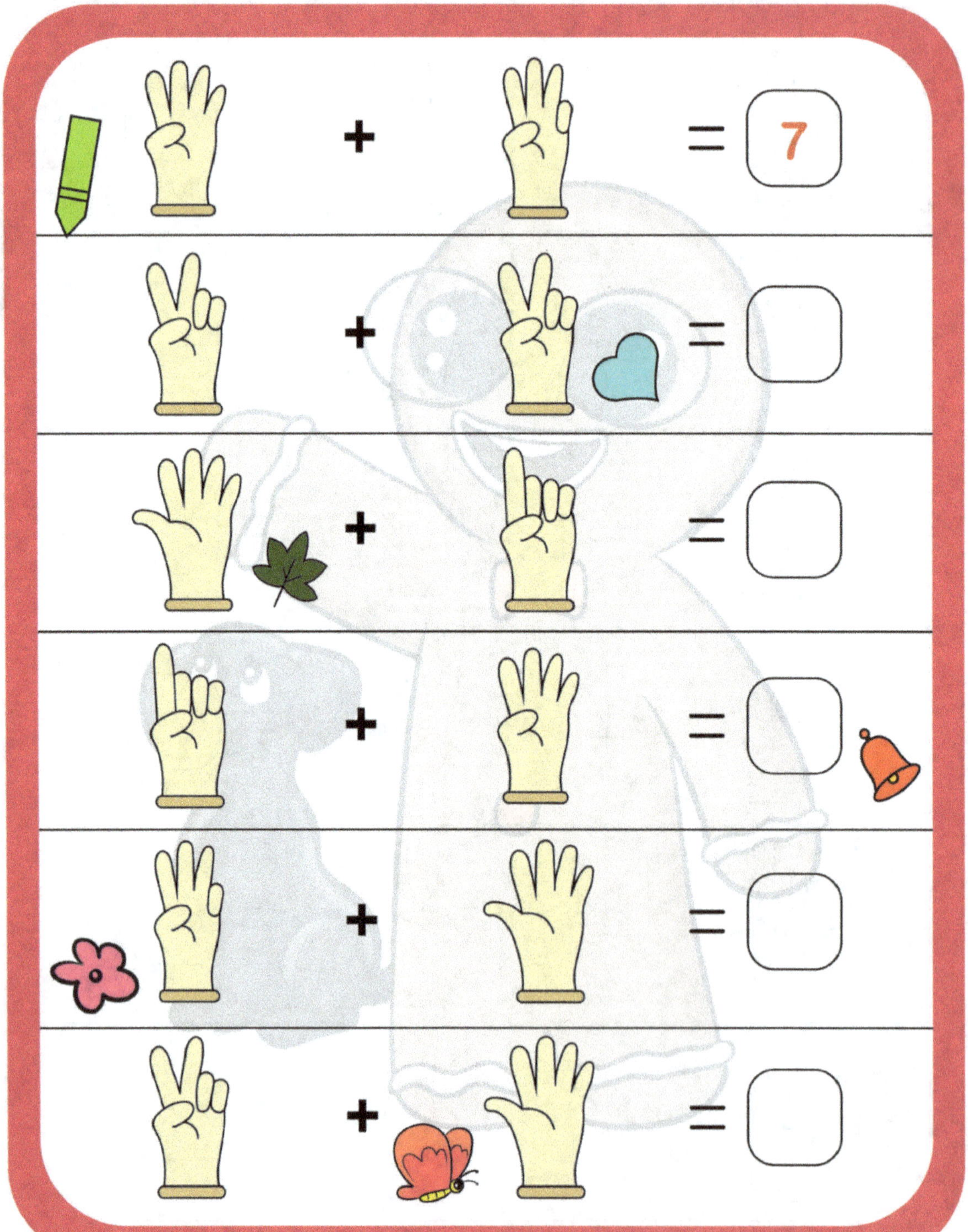

4 + 3 = **7**

3 + 3 =

5 + 2 =

2 + 5 =

4 + 5 =

3 + 5 =

 # Fill in the Blanks

_____

1	O___**n**___e	n u e
2	_____wo	W M T
3	Th_____ee	r c v
4	Fou_____	u a r
5	F_____ve	e i q
6	S_____x	n i y
7	Se_____en	v w o
8	Eig_____t	h r g
9	Ni_____e	n p u
10	_____en	T A N

Fill in the Blanks

_____

11	Ele____v____en	v	o	e
12	Twe_____ve	l	r	i
13	Thi_____teen	r	h	w
14	Four_____een	t	q	r
15	Fi_____teen	r	f	q
16	Si_____teen	t	i	x
17	Se_____enteen	v	f	b
18	Eig_____teen	z	h	g
19	Nin_____een	t	j	u
20	T_____enty	w	r	g

Fill in the Blanks

$$
\begin{array}{cc}
2 & 7 \\
\times & \boxed{3} \\
\hline
8 & 1
\end{array}
$$

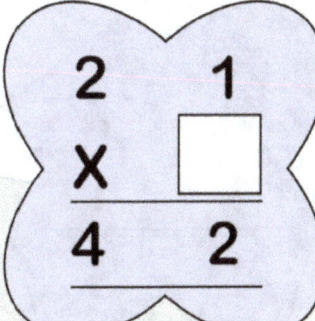

$$
\begin{array}{cc}
2 & 1 \\
\times & \square \\
\hline
4 & 2
\end{array}
$$

$$
\begin{array}{ccc}
4 & & 1 \\
\times & \square & \\
\hline
2 & 0 & 5
\end{array}
$$

$$
\begin{array}{ccc}
6 & & 8 \\
\times & \square & \\
\hline
1 & 3 & 6
\end{array}
$$

$$
\begin{array}{ccc}
3 & & 9 \\
\times & \square & \\
\hline
2 & 3 & 4
\end{array}
$$

Fill in the Blanks

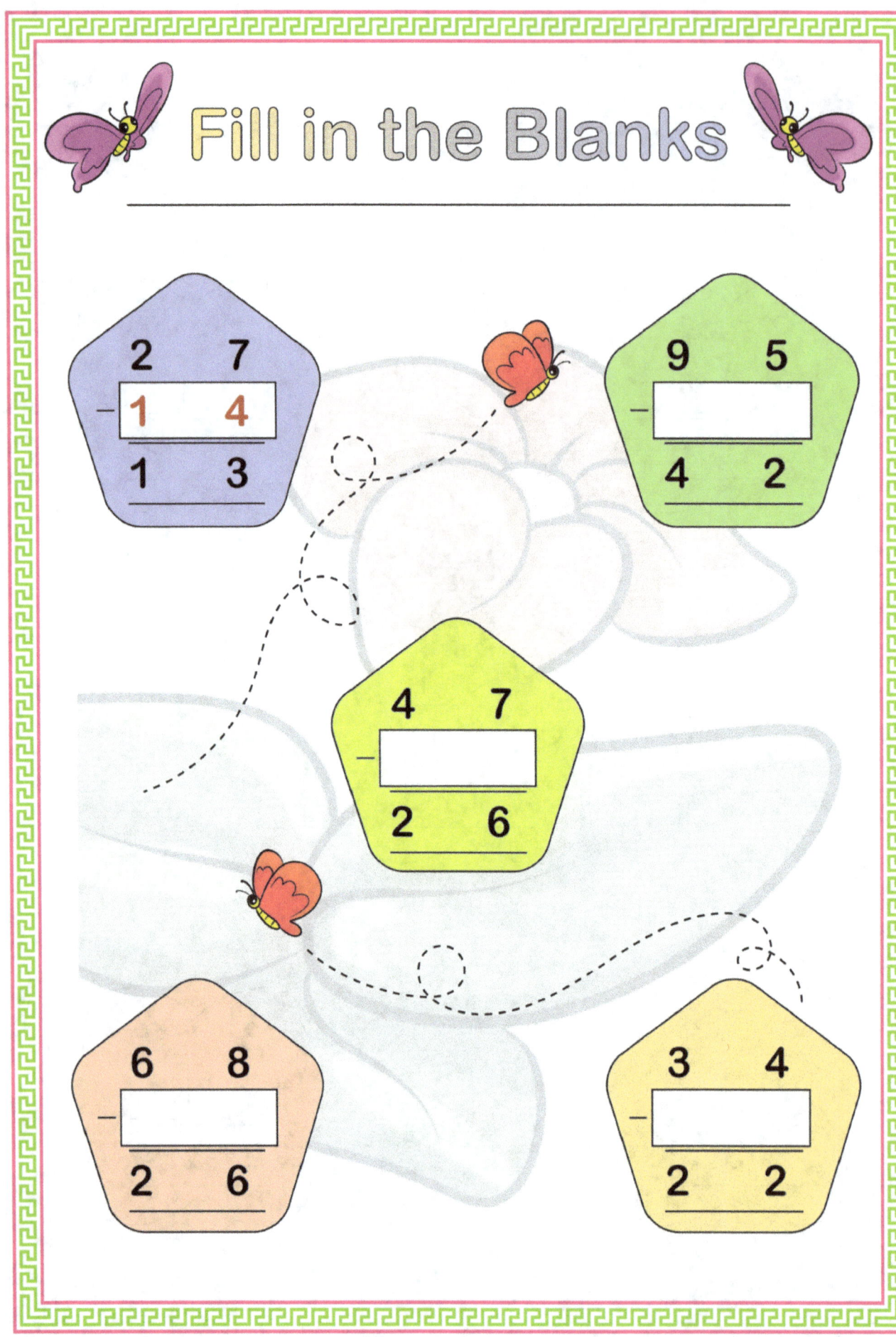

$$\begin{array}{r} 2\ 7 \\ -\ 1\ 4 \\ \hline 1\ 3 \end{array}$$

$$\begin{array}{r} 9\ 5 \\ -\ \ \\ \hline 4\ 2 \end{array}$$

$$\begin{array}{r} 4\ 7 \\ -\ \ \\ \hline 2\ 6 \end{array}$$

$$\begin{array}{r} 6\ 8 \\ -\ \ \\ \hline 2\ 6 \end{array}$$

$$\begin{array}{r} 3\ 4 \\ -\ \ \\ \hline 2\ 2 \end{array}$$

Find the Way

Hint
Follow 7

1	0	3	4	5	3	5	2	9	6
7	7	2	4	7	7	7	8	4	9
1	7	7	7	7	6	7	7	9	5
3	2	6	4	5	7	7	7	0	4
1	2	3	4	7	7	3	8	9	6
1	7	7	7	7	6	1	8	7	3
7	7	3	1	5	6	0	1	9	8
7	2	9	4	5	6	7	7	7	1
7	7	3	7	7	7	7	8	7	2
1	7	7	7	5	6	2	3	7	8

Make Division & Color it

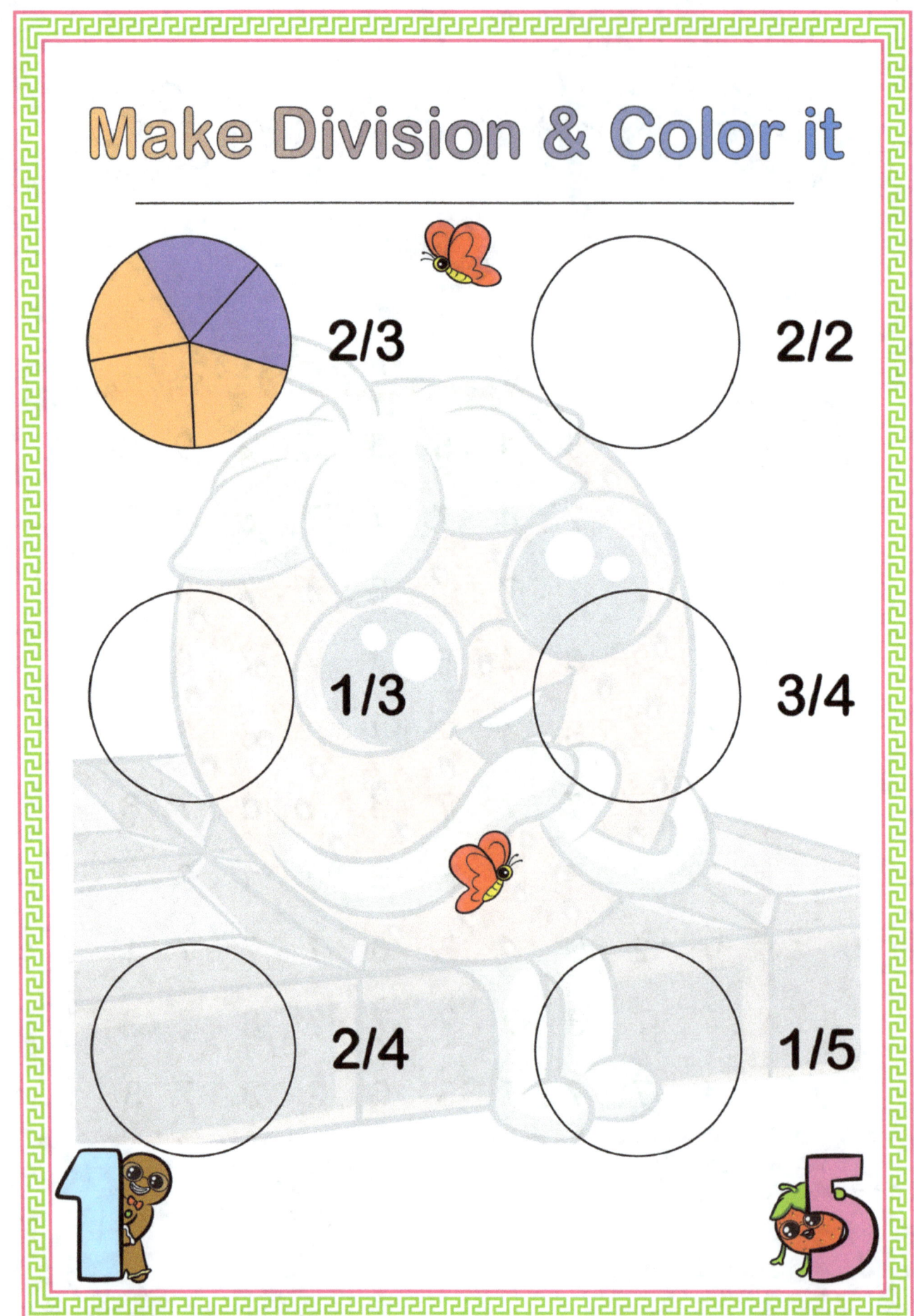

2/3

2/2

1/3

3/4

2/4

1/5